Live Music!

Brass

Elizabeth Sharma

Wayland

Titles in the series

Brass	Strings
Keyboards	The Voice
Percussion	Woodwind

TOPIC CHART	MUSIC			Science	Maths	English	Geography	History	Technology
	Performing and Composing	Knowledge and Understanding							
Chapter 1 What are brass instruments?		✓				✓	✓	✓	
Chapter 2 Making brass sounds	✓	✓		✓		✓			✓
Chapter 3 Brass in the past		✓		✓		✓		✓	
Chapter 4 Blow your own trumpet	✓	✓			✓	✓			✓

First published in 1992 by
Wayland (Publishers) Ltd
61 Western Road, Hove
East Sussex BN3 1JD, England

Editor: Cath Senker
Designer: Malcolm Walker
Consultant: Valerie Davies, Primary Adviser,
East Sussex County Council Music School

British Library Cataloguing in Publication Data
Sharma, Elizabeth
   Brass. – (Live music! series)
   I. Title II. Series
   788.9

ISBN 0 7502 0449 4

Typeset by Kudos Editorial and Design Services,
Sussex, England
Cover artwork by Malcolm Walker
Printed and bound by Casterman S.A., Belgium

Contents

Live Music!

What are brass instruments?

Have you ever heard a **fanfare**? The bright, thrilling sound of the brass instruments announces that something important is about to happen.

Brass instruments are not always made from brass. But they are all made from a long tube opening out into a bell at the end. The player blows directly into the instrument, usually through a mouthpiece.

Brass instruments sound loud and clear. They can be heard from far away. From ancient times until early this century, they were used to signal instructions to troops in battle.

This American girl is playing a fanfare on the bugle. A bugle is a trumpet without valves.

4

Didgeridoos and alpenhorns

In many parts of the world, trumpets and horns are played to make a loud, sometimes **rhythmic** sound. Often they are used for celebrations.

The Australian Aborigines play the didgeridoo, which is a long, beautifully carved wooden tube, made from a **eucalyptus** branch. It is blown to produce a rhythmic **drone**.

This is a didgeridoo. It is a brass instrument because the player blows directly into it to make the sound.

The alpenhorn from Switzerland is also made from a large hollowed-out log. It was originally made for signalling from one mountain top to another.

These Swiss men are playing alpenhorns. Very simple melodies can be played on the alpenhorn.

In Tibet, another mountainous country, Buddhist monks play instruments rather like alpenhorns, that make a booming sound. They are played during religious ceremonies, or to mark the start of festivals.

Orchestral brass

These trombone players are practising before a concert.

The brass section of the orchestra usually plays when the music needs to be loud and exciting. The brass instruments in a **symphony orchestra** are the trumpets, French horns, trombones and tuba.

This girl is playing the tuba in an orchestra.

Here is a French horn player. His right hand is inside the bell. This helps him to support the instrument. It also helps him to alter the sounds that come out.

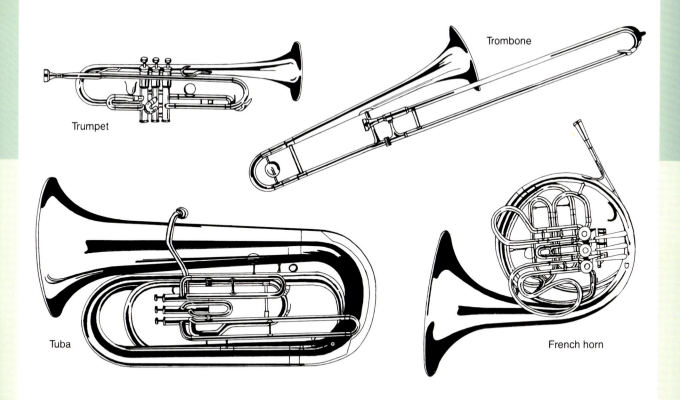

Trumpet

Trombone

Tuba

French horn

The trumpet has the highest **pitch** of all the orchestral brass instruments. The French horn is lower-pitched. French horns can be played loudly, but can also sound soft and mellow. They blend well with woodwind instruments, as well as with the rest of the brass section.

There are usually two **tenor** trombones and one **bass** trombone in the orchestra. The tuba is the lowest-pitched brass instrument. It can produce very low notes indeed. Although it is a large, heavy instrument, it can be played softly.

Here are the orchestral brass instruments. The trombone in the picture is a tenor trombone. The bass trombone is almost the same, but it has an extra length of tubing.

Marching brass

There are many brass instruments in marching bands and **military bands**. Their brilliant sounds can be heard from far away and people flock to hear them.

Marching bands are very popular in the United States. Many colleges have huge bands which perform complicated marching routines while playing music.

This marching band in San Antonio, USA, is playing at a fiesta (festival). Can you see the trombone players? The small, silver-coloured instruments like trumpets are called cornets.

There are several instruments which are used in brass bands but not in orchestras. The cornet is similar to the trumpet, with a softer, less shrill sound. Cornet players usually play the tune. The tenor horn, **baritone** horn and euphonium look like small tubas. Their mellow tones blend well together.

Can you see the tubas on the right of the picture? In front of them are the tenor horns. The baritone horns are in front of the trombones.

Sousa's ready-to-wear tuba

This man is playing a sousaphone in a marching band. Some sousaphones are made from a light material called fibreglass. They are even easier to carry than this metal one.

The American composer, John Philip Sousa (1854–1932), wrote a lot of music for marching bands. He invented the sousaphone. It was similar to the tuba, but could be worn around the player's body. This made it much easier to carry than the tuba when marching.

Making brass sounds

Brass instruments are blown in a special way. Make a circle with your thumb and forefinger and put it against your lips. Now, keeping your lips closed, blow hard. What happens? Are you making a buzzing sound?

The air you are blowing makes your lips **vibrate** very fast. This is how you have to blow to produce a sound on a brass instrument. The harder you blow, the higher the buzzing sound will be, as your lips vibrate faster.

Benezie, Rita and Fabio are making a buzzing sound through their lips. They are being careful not to puff out their cheeks. Brass players have to learn to make a smooth, buzzing sound. Then they will be able to produce musical notes on their instruments.

Now try putting your lips to the opening of an empty plastic drink bottle. Can you make a sound? Ask an adult to cut the bottom off the plastic bottle, and try again. Can you produce a sound now? What have you learnt from this?

This boy is trying to blow a note through a plastic bottle with the end cut off.

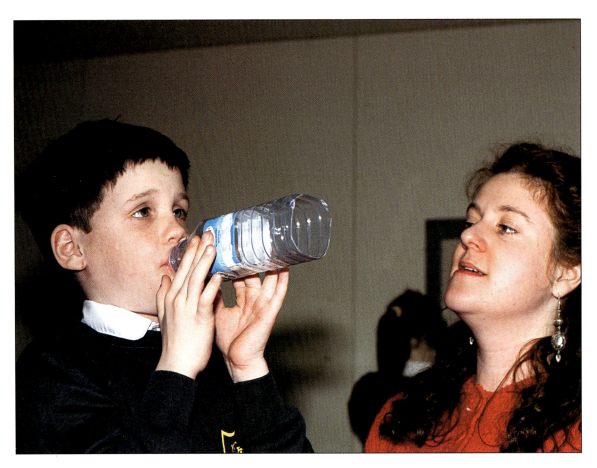

Cups, tubes and bells

Brass players rest their lips against a cup-shaped mouthpiece. The mouthpiece fits into the tube of the instrument, which opens out into a bell at the end.

Compare the sounds of the trumpet and the tuba. The trumpet has a short length of narrow tubing. It makes high, bright, clear sounds. The largest tuba has 14 m of wider tubing coiled up. It makes a low, booming sound.

The longer and wider the tube, the lower the sound. The design of the mouthpiece, and the shape of the tube, give each brass instrument its own special sound.

These boys are playing tubas in a band. Notice the large mouthpieces on the instruments. It is not hard to blow a note on a tuba.

Lips and valves

Simple brass instruments are just tubes with mouthpieces. The players make different notes by blowing harder or softer, and by changing their lip position within the mouthpiece.

Try blowing a brass instrument or a plastic bottle with the bottom cut off. Buzz your lips into the mouthpiece and try to produce a smooth sound.

Then relax your lips a little, blow more softly, and find the lowest note you can play. Press your lips more tightly together and blow harder to produce a higher note. See how many different notes you can play, and then find them on the piano.

Sharif is pressing his lips tightly together and blowing hard. He is playing a high note on his cornet. Katherine is relaxing her lips and blowing softly. She is producing a low note on the tenor horn.

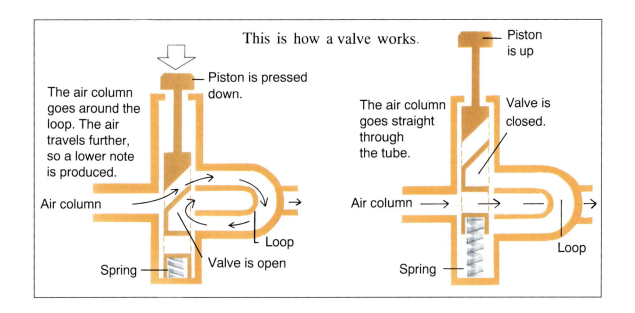

This is how a valve works.

Piston is pressed down.

The air column goes around the loop. The air travels further, so a lower note is produced.

Air column

Spring

Valve is open

Loop

Piston is up

The air column goes straight through the tube.

Valve is closed.

Air column

Spring

Loop

Many brass instruments have valves, which open and close different lengths of tubing connected to the instrument. The valves allow the musician to play all the notes of the **scale**.

The trombone (right) has a slide which can lengthen or shorten the tube to make different notes.

Brass in the past

Here is a monk in Darjeeling, India, blowing a horn. This horn is very simple, like the ones played in ancient times.

In ancient times, people found that they could produce a loud sound by blowing into hollow animal horns, large shells, or carved wooden horns. All over the world, these instruments were used for signalling during battle and hunting. Similar instruments are still used in many parts of the world.

When people learnt to work with metals, trumpets and horns were made from bronze and silver. A silver trumpet was found in the tomb of the Egyptian king, Tutankhamun, who died in 1323 BC.

All through history, brass instruments have been played outdoors in parades and ceremonies. They are used to announce the arrival of someone important, like a queen.

This picture shows a stone carving of Roman musicians in a procession. It was carved in AD113. Can you see the rounded horns being played?

Brass joins the orchestra

Brass instruments were used in the early orchestras. They were usually played when special effects, or loud sounds were needed.

In the sixteenth and seventeenth centuries, trumpets were mainly used for playing **solos**. These trumpets had no valves, so the player had to form all the notes using only the lips and breath pressure.

This picture shows German musicians in 1520. Look at the man fifth from the right. He is playing a simple trumpet without valves.

A trumpet player could not produce all the different notes of the scale. To play the high notes, the trumpeter had to blow hard and play loudly. Only very skilled musicians were able to make a good sound.

During the eighteenth century trumpets, horns and trombones were often played in orchestras.

At that time, European instrument-makers invented extra sections of tubing for brass instruments. They were called crooks. The crooks could be added to lengthen the tube and produce different sets of notes.

The players needed time to change the crooks, so composers had to write music with lots of **rests** in the brass parts!

In 1815, piston valves were invented to open and close different lengths of tubing connected to the instrument, so the awkward crooks were no longer needed.

This is an eighteenth-century French horn with crooks. Can you see how the crooks are joined on to make the tube longer?

The orchestra grows

After valves were invented, composers could give the brass section more interesting parts to play. Brass instruments with valves could produce a complete range of notes. They could play softly as well as loudly.

This is a picture of an English concert in 1849. Can you pick out the brass instruments near the back of the orchestra?

Composers began to write music for a larger number of brass instruments, and so more woodwind and stringed instruments were needed to balance their loud sound. During the nineteenth and twentieth centuries, orchestras grew even larger.

Some composers, such as Antonin Dvořák (1841–1904) and Modest Mussorgsky (1839–1881) wrote descriptive music for these huge orchestras. This was music that used the many different instruments to paint pictures, or tell stories, using sound.

Here is a portrait of the composer Antonin Dvořák, who came from Bohemia (which became part of Czechoslovakia). He wrote beautiful descriptive music, often using folk songs in his works.

Nowadays some orchestras have nearly one hundred musicians. Imagine being a member of a huge orchestra like that, surrounded by all the different sounds! The brass section plays at the most exciting moments – sometimes **chords** and sometimes fanfares.

The lower-pitched instruments, the tuba and bass trombone, may also play frightening sounds. The horns may play soft, smooth music, like the passage in Mendelssohn's *A Midsummer Night's Dream*.

Brass bands

Brass music gives a feeling of excitement and pride. Throughout history, young men have been encouraged to join the army after following a military band.

From the end of the eighteenth century, English companies began to form brass bands for their workers. This became a popular form of **amateur** music-making, particularly in the north of England.

This is a brass band in the north of England. The man in front is carrying a euphonium. He is wearing old-fashioned clothes. A band player in the nineteenth century might have worn clothes like these.

Many local coal-mines had their own bands, and the local people felt very proud when their bands won music competitions. Although many mines have closed, the brass bands still continue.

Jazz music

Jazz music began among the black people of the southern states of the USA, at the end of the nineteenth century. From the beginning of jazz music, the trumpet and trombone were very popular instruments. The bright, rhythmic sound of the trumpet particularly suited jazz **melodies**.

This is Louis Armstrong, the famous American jazz trumpeter who became popular in the 1920s. He played brilliant trumpet solos.

In the 1920s, a jazz band usually had a trumpet, a trombone, a clarinet, drums and a guitar. The musicians often played **improvisations**. They made up the music as they went along.

Brass instruments, mainly trumpets and trombones, were played in the **swing** bands of the 1930s and 1940s. These bands, such as the Count Basie Band and the Duke Ellington Band, played to large audiences in dance halls around the USA.

This is a swing band, playing in the USA in 1940. Swing bands combined groups of trumpets, trombones and saxophones. There would also be a double-bass, a piano, and sometimes even a guitar.

Jazz rhythms and tunes found their way into many other kinds of music – even classical music. Jazz is still popular today. Jazz styles are often mixed with other kinds of music to make lively dance music. Listen to some different kinds of jazz and compare how the instruments are used.

This is the well-known South African trumpeter, Hugh Masekela. His band combines jazz melodies with African musical styles and rhythms.

Live Music! Blow your own trumpet

These girls are joining a cardboard tube and some plumbers' piping together to make a long trumpet.

Plastic bottle.

Plastic plumbers' piping fits into plastic bottle.

Top part of a plastic bottle.

Look around and find a tube which you could use to make your own trumpet. You could use the cardboard tube from a roll of kitchen foil or wrapping paper, a length of plastic plumbers' piping, or a length of hose-pipe.

Ask an adult to cut the bottom off a large plastic bottle for you. Attach the bottle to one end of your tube to act as a bell. You may need to use sticky-tape or modelling clay to attach it. You could use the top part of a plastic bottle, where the cap screws on, to make a mouthpiece. Ask an adult to cut it for you.

See how many different notes you can blow on your trumpet. Now make up a fanfare, using a recorder, xylophone or piano. Remember, you can only use these notes:

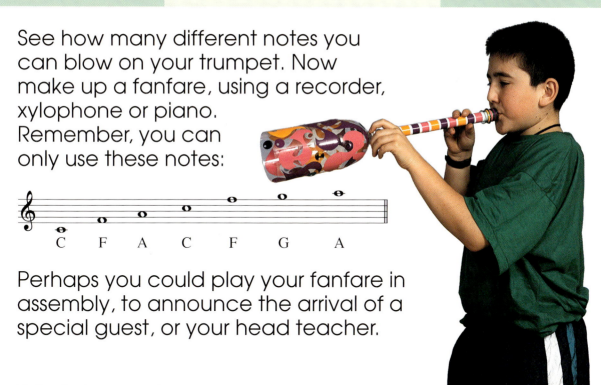

Perhaps you could play your fanfare in assembly, to announce the arrival of a special guest, or your head teacher.

Lights out!

Here is a short piece of bugle music. It was first used by the British Army as a signal for 'lights out' at the end of the day:

Make up some short tunes which you could use to signal to your classmates at certain times of the day, such as:

Registration   Lunch time

Assembly   Home time

Glossary

Amateur   Someone who does an activity for fun rather than as a job.

Baritone   The second lowest-pitched instrument in a family of instruments.

Bass   The lowest-pitched instrument in a family of instruments.

Chord   A group of notes which are played together to produce an agreeable sound.

Drone   A bass note or chord played over and over again to accompany a melody.

Eucalyptus   A type of tree often found in Australia.

Fanfare   A short tune played on brass instruments, used as a signal at special events.

Improvisations   Music that is made up by the player as he or she is performing.

Melodies   Tunes.

Military bands   Army, navy or air force bands, usually made up of brass, woodwind and percussion instruments.

Pitch   How high or low a note sounds.

Rests   Periods of silence in a musical piece.

Rhythmic   Having a regular beat.

Scale   The seven different notes in Western music (known by the letter names A,B,C,D,E,F,G and the half-notes between them).

Solos   Parts of a piece of music that are written for just one instrument.

Swing   A kind of jazz music that was popular in the 1930s and 1940s.

Symphony orchestra   A large orchestra with all the instruments needed to play symphonies – brass, woodwind, strings, harp and percussion.

Tenor   The third lowest-pitched instrument in a family of instruments. The baritone and bass are lower.

Vibrate   To shake very quickly.

Finding out more

1. Why not listen to some brass music?

Descriptive orchestral music: The last movement of *Symphony No. 9 (From the New World)* by Dvořák
Pictures at an Exhibition by Mussorgsky

Orchestral brass: *Praise be to the God of Brass* from *Belshazzar's Feast* by William Walton

French horn: The horn solo in the second movement of Tchaikovsky's *Fifth Symphony*

Jazz: Count Basie Band, the Duke Ellington Band, Louis Armstrong

Trumpet: The trumpet solo in Bach's second *Brandenburg Concerto*

Tuba: *Tubby the Tuba* by George Kleinsinger, in *Hello Children Everywhere*, Vol. I, EMI

Trumpet and tuba: *The Young Person's Guide to the Orchestra* by Benjamin Britten

2. Try to hear some live music. Look out for marching bands and brass band concerts in the park. Go to a concert given by a local youth band or orchestra. You can find out about the performances from your local library.

Useful books

Brass by Dee Lillegard (Childrens Press, 1988)
The Eyewitness Guide to Music ed. Janice Laycock (Dorling Kindersley, 1989)
The Orchestra by Mark Rubin (Oxford University Press, 1989)
Twentieth Century Music by Alan Blackwood (Wayland, 1989)

Index

Page numbers in **bold** indicate subjects shown in pictures as well as in the text.

Acknowledgements
The photographs in this book were provided by: Birmingham City Council 7 (above); Cephas (N.Blythe) 6 (below); Chapel Studios 7 (below), 12, 15; C M Dixon 19; East Sussex County Council 8; Eye Ubiquitous (D.Fobister) 24; Fotomas Index 21; Hulton Picture Library 26; Life File (J.Hoare) 17; Mary Evans Picture Library 20, 22, 23; Photri (B.Kulik) 4, 6 (above),18; Rex Features 27; Topham 5, (A.Carey) 10, 25; Wayland Picture Library (Isabel Lilly) 14, (all Garry Fry Stills Photography) *cover*,13, 16, 28, 29; ZEFA (J.Flowerdew) 11. Artwork: Creative Hands 17, 28; Malcolm Walker 9.
The publishers would like to thank the staff and pupils of the Hammersmith School, London, for their kind co-operation.